YOUR KNOWLEDGE HAS VALUE

- We will publish your bachelor's and master's thesis, essays and papers

- Your own eBook and book - sold worldwide in all relevant shops

- Earn money with each sale

Upload your text at www.GRIN.com and publish for free

Bibliographic information published by the German National Library:

The German National Library lists this publication in the National Bibliography; detailed bibliographic data are available on the Internet at http://dnb.dnb.de .

Imprint:

Copyright © 2019 GRIN Verlag
Print and binding: Books on Demand GmbH, Norderstedt Germany
ISBN: 9783668949041

This book at GRIN:

https://www.grin.com/document/465422

Dr. Roshan Adhikari

Heater at Less Cost, Rather More Benefit. A Concept Proposal

GRIN Verlag

Heater at Less Cost, Rather More Benefit: A Concept Proposal

Prof. Roshan Adhikari

Research Presented to Grin Publishing, Germany

2019

Table of Contents

1. Background

The carbon burning is a practice which is attributive to global warming and similarly the solar heaters are also affecting the environment with harmful means to body organs because of charged sub-atomic particles. The burning fire is harmful to body in that it can cause suffocation and respiratory problems, while electric heaters and ACs are similarly health problematic to bring dehydration of body and the disease of respiratory system like asthma. The demand of the doctors and heat basking consumers is the kind of heating system which means to be without dehydrating, blood clotting and suffocating effects on body. For this purpose a self-generation of heat energy is required which should not need fire, electricity, chemical reaction process, rather than the physical change mechanism had nothing to do with a pollution of environmental construct. The design of such a prospective facility for both health and global milieu has been made feasible after four year serial researches of this present author.

The heater had without the use of fire and electricity may be manufactured by virtue of exothermic reaction between inorganic chemicals which will pollute respiratory system through surrounding atmosphere of a room, besides its very expensiveness. A flameless ration heater, or FRH, according to general report from Administration, is a water-activated exothermic chemical heater included with meals, ready-to-eat, that is used to heat the food. US military specifications for such an easily portable heater require the same as being capable of raising the temperature of an eight-ounce (226.8 g) entree by 100 °F (56 °C) in twelve minutes, and that it has no visible flame.

The ration heater useful for both cooking food and making the body warmer contains finely powdered inorganic element magnesium, alloyed with a small amount of iron, and table salt. To activate the reaction, only a small amount of water is to be added, and the boiling point of this same liquid is quickly reached as the reaction proceeds.

Ration heaters generate heat in an electron-transfer process called an oxidation-reduction reaction because of the friction of charged elementary particles with molecules of metallic layer and the evolved heat is flowed to different required sites for several uses. Water oxidizes magnesium metal, according to the following chemical reaction:

$$Mg + 2H_2O \rightarrow Mg(OH)_2 + H_2 \ [+ \ heat \ (q)]$$

This reaction is analogous to iron being rusted by oxygen, and proceeds at about the same slow rate, which is too slow to generate usable heat. To accelerate the reaction, metallic iron particles and table salt (NaCl) are mixed with the magnesium particles.

Iron and magnesium metals, when suspended in an electrolyte, form a galvanic cell that can generate electricity. When water is added to a ration heater, it dissolves the salt to form a salt-water electrolyte, thereby turning each particle of magnesium and iron into a tiny battery. Because the magnesium and iron particles are in contact, they become thousands of tiny short-circuited batteries, which quickly burn out, producing heat in a process the patent holders call supercorroding galvanic cells.

One brand of self-heating rations uses 7.5 grams of a powdered magnesium-iron alloy, consisting of 95 % magnesium and 5 % iron by weight, 0.5 grams of salt, in addition to an inner filler and anti-foaming agent. Upon adding 30 milliliters of water, this mixture can heat a 230 gram meal packet by 100 °F in about 10 minutes, releasing approximately 50 kilojoules of heat energy at about 80 watts (Administration, 2008-2009).

As per Zen Adventuring Blog, these said types of magnesium heater are being used in carrying out a purpose with canned foods, Heater Meals, the Mountain House's Mountain Oven, Hot Pack, rations for space administration and US military and paramilitary meals. They can provide a simple as well as safe method for heating food, with little risk of fire or explosion.

The most popular design uses heating pads made from a super corrodible magnesium and iron alloy in a porous matrix formed from polymeric powders with sodium chloride incorporated with it or in a separate tablet. To heat, water is added either from an external source or by puncturing a container of water onto the heating pad. The water dissolves the sodium chloride into an electrolyte solution causing magnesium and iron to function as an anode and cathode, respectively. An exothermic oxidation-reduction reaction between the magnesium-iron alloy and water to produce magnesium hydroxide, hydrogen gas and heat (zenstoves.net, 200-2009). The much preferable magnesium heater is though safe from the case of accident due to higher chemical reactivity but still it is risky to handle magnesium because water drops fallen on the same electropositive metal with the liberation of its hydrogen by the former will suddenly cause an outburst of fire due to recombination of the gas with atmospheric oxygen all as sudden.

2. Problem Statement

Directly basked sun will give health benefits more than through solar powered equipment. Sun emitted rays can cause vitamin D to form in the bones while photovoltaic cell used to convert solar energy into electricity and this used to produce current in coils may instead emit charged electrons in the close to atmosphere. The flow out of electric charges is responsible for cellular damage and dryness of living being in an environment. Consequently the body will have water deficiency in blood due to solar battery released waves of electricity from a such type powered device which carry energetically vibrating particles released from a certain potential configuration within the atomic shell. These unlike the photons of light which can emit electrons from matter have electrification to the surrounding bodies and living organisms. Yet no system of device has been improvised which can have use of heat for warmth without a threat about global warming after a soaking effect to blood, cell and onset of respiratory disorder in human expectants the present energy sources give everywhere (Adhikari, 2018).

3. Theory

The theory for the safe use of abundant resources for universal energy is based on light deflection by magnetic field and convergence of rays into raised temperature of heat. According to the idea of magnetic field effect the propagating beam of charged particles electrons moving in a magnetic field will be deflected by that field at a right angle to both the field and to the direction of the particle. As is applied that rule, the electrons in the cathode ray are travelling in opposite *to* the flow of conventional current. In addition to experimenting with magnets, scientists also experimented to if charged plates are positioned near the cathode ray tube the results will show electrostatic deflection as opposed to the electromagnetic deflection. The electrons in the cathode rays would deflect toward the positively charged plates, and away from the negatively charged plates. This a rare incidence of bending of light may occur under the influence of a powerful magnetic field since light is the beam of photons in the oscillating electromagnetic waves toward a transverse direction and can be deflected easily by gravity from massive bodies in space. If one goes to the photon level, firing individual photons through the opening of the capacitor plates: photons as neutral will change direction only if they interact. There exists a small probability that a photon meeting an electric field may interact with the electromagnetic interaction, but the probability is tiny as the relevant Feynman diagram has at least five coupling constants (Pawar, 2017).

4. Objective of the research

The research is mainly aimed at the usage of heat energy along-with lower cost, care about environmental harms and unwanted states of health aberration.

5. Discussion

Glycerin is derived from fat of animals and plants as a compound said to be glycerol. It vaporized soon with higher temperature when heated to a required degree. More than that temperature level it will become carcinogenic if further heat is given. The hot glycerin mixed with plain and normal water inside a so bigger bottle of glass is useful to have radiation of heat outside to the room when the weather will be cold for need of warmth. As the duration of temperature remaining in glycerin is very short a small coil filament should be fitted to the bottom of bottle charged by a dry cell. The supply of voltage can last long period from the battery to coil and will maintain the required temperature continuously of glycerin mixed with water in the bottle by a heat transfer. In the way the warmth to be availed from hot glycerin can be made possible to spread around the cold room through refraction by water and glass until the small battery ends, perhaps not less than ten days (Adhikari, 2018).

If it was not safe it would not be used as the base for vape juices because vaping is designed to be a safe alternative to smoking. Remember a safe alternative, not a safer alternative. The exhaled vapor from vaping is safe for anyone who breathes it in, except maybe for the rare few with allergies to glycerin. There are no second hand dangers for children and it does not stain household fabric (furniture, curtains etc.) with a lasting foul odor whilst vaping there is no danger of heating glycerin to the required temperatures for it to become carcinogenic, only dry hits can cause this but it is from the wicking materials and not the glycerin itself, besides the taste is so foul and so quickly detectable that you cannot inhale enough for it to do any damage that about 2–3 puffs of a cigarette cannot do (HashFish, 2018).

The above type of simple electrifiable device can perform for short time depending on the resistivity of coil and potential of D.C. source therefore a durable management for the emission of heat from light can be thought with the application of some other chemical as well as physical means having some few change processes of energy forms. Auerbach explains light from the sun excites electrons in the atoms which constitute the brick wall. How that electronic energy get converts to heat, any one may ask. The key is radiation less transitions. The atoms of the brick are perpetually vibrating. Some of those atoms vibrate sufficiently vigorously that their vibrational energy is roughly equal to the electronic energy (photons) absorbed from the sun in essence, they are in resonance with the solar energy. Those atoms then make a quantum transition from electronically excited to vibrational excited, meaning that the energy causes the whole atom to move. One feels that motion as heat. The atoms which make the jump to vibrational excitation soon collide into neighboring atoms, dissipating their vibrational energy throughout the entire brick, making the brick hot throughout (Auerbach, 2018). The heat from solar system is derived through the photovoltaic conversion of white light into electricity at first then again the further change into that radiant energy through passage of current into loops of wire in an oven .

The sun produced heat is good for health rather than artificially converted means of radiation for an indoor temperature gain. That is directly basked sun will give health benefits more than through solar powered equipment. Sun emitted rays can cause vitamin D to form in the bones while photovoltaic cell used to convert solar energy into electricity and this used to produce current in coils may instead emit charged electrons in the close to atmosphere. The flow out of electric charges is responsible for cellular damage and dryness of living being in an environment. Consequently the body will have water deficiency in blood due to solar battery released waves of electricity from a such type powered device which carry energetically vibrating particles released from a certain potential configuration within the atomic shell. These unlike the photons of light which can emit electrons from matter have electrification to the surrounding bodies and living organisms (Adhikari, 2018).

In a big jar of glass normal water is to be poured half of the full volume. In the water one of the hydrocarbons e.g. petrol, alcohol, xylene, toluene, etc.is to be mixed. Beneath the same glass container should be placed a strong magnet but is better the flat and larger in size to occupy the whole bottom of the container. In the day time otherwise during night when a bulb is lit the light is deflected by the magnet toward the bottom of jar and make converge the rays in a concentration depending how much the magnet is powerful. The rays of light have temperature because light is a radiation with frequency of oscillation about wave like particles where the change of energy of vibration can give rise to a heat form. As rays of light are converged therein at the point of glass container's bottom due to an attraction of magnet the temperature out of radiated heat derived from soft light inside a room burns slowly the already inflammable chemicals like petrol, diesel, xylene, toluene, alcohol with warmth. The mechanism is useful to heat up a cold environment since there is no danger of fire because of heating hydrocarbon with the energy of light. On the one hand light produced heat is not intense to blaze the said chemical substance, on the other the coldness of water in the glass bottle acts as a coolant to soothe the flame when highly powerful magnet happens to cause fire in sunlight for the chemical fuel mixed in water, otherwise magnifying glass should not be used with the bottle to heat the liquid which concentrates excessive light and evolves greater temperature that will definitely ignite same given mixture with so an inflammable heat.

6. Experimentation

After such a household invention of the present author done, a medium sized bottle was filled with normal water up-to more than its half of the level while showing demonstration about scientific and technological research by self in office, Nepal Philosophical Research Center, Kathmandu on 2019-january-12. The remaining volume of the bottle was poured in with petroleum liquid. Then further 20 drops of rectified spirit were added in order to increase the concentration of hydrocarbon liquid and consequently its inflammability. After that again a sealed bottle of pure glycerin was dipped into the so prepared mixture. The cap of the bottle containing all necessary materials had been closed. The bottle was placed over a powerful magnet and around the glass wall of the same container other pieces of magnet were wrapped for increasing the attractive power of these to deflect light to their possible capacity. Later a few moments the bottle gave warmth to surrounding environment of room. The rectified spirit could be added further after two hours when the heat temperature reduces because of the usage of inflammable substances like petrol and spirit in the water mixture (Adhikari, 2018).

7. Conclusion

The rays of light are attracted though to rare extent by magnets as some convergent point of mixture. The converged rays will grow heat and temperature which is will be captured by glycerin of bottle placed inside the mixture. Glycerin immediately increases the temperature with little heat gained from the solution which radiates into the hydrocarbon of bigger bottle. The hydrocarbon fuel again burns because of addition heat gained from glycerin. The increase temperature of the same mixture is absorbed by the glycerin and this process recycles until the heat is not finished after being fully radiated to outside through the glass wall. The hydrocarbon fuel should not burn with inflammation, for this purpose water in the bottle acts as a stabilizer of the temperature gained due to light and also heat recycle (Adhikari, 2018).

8. Personal Recommendation

To increase the temperature of devised heating system for use the quantity of water is reduced, the same of hydrocarbons increased, wherein strongly inflammable substance benzene can be used instead of other rectified spirit, petroleum, xylene, toluene, etc. like hydrocarbons; a magnifying glass is preferred to install into the little hole made up through the bottle cap so a red light emitted from led bulb of a torch attached to may directly be incident upon the inner solution across it. The day light of room is enough for giving some temperature to the prepared mixture while red electric bulb can further be used for the purpose of gaining intense heat when needed during day and overnight periods simultaneously with convergence of light rays by role super-magnet placed below that bottle. In case of an increased higher temperature to be possibly happened there the bottle may earlier be chosen of hard glass so not to have danger of explosion from hydrocarbon mixture. In total the cost of the required materials may reach up-to US$ 10 to get prepared a heater capable of giving warmth to any big and non-partitioned venue without the use of reactive chemicals, fire and electricity. The process of gaining heat out of light being focused with the help of powerful magnet on inflammable organic chemical and same energy recycled by glycerin as absorbent is usable to radiate room walls through pipes wherein so prepared hot liquid can be supplied all around from any big tank of tall house (Adhikari, 2018).

Bibliography

Administration, F. A. (2008-2009). Flameless ration heater. *Wikipedia, the free encyclopedia: Jump to navigation, Jump to search* , 1.

Auerbach, S. M. (2018). How exactly does light transform into heat--for instance, when sunlight warms up a brick wall? I understand that electrons in the atoms in the wall absorb the light, but how does that absorbed sunlight turn into thermal energy? *Scientific American* , 1-4.

HashFish. (2018). Is it safe to heat glycerin? *Quora* , 1.

Pawar, K. (2017, April 13). Is light deflected by external electric and magnetic field? *Physics* , pp. 1-3.

zenstoves.net. (200-2009). *Flameless Chemical Heaters.* Zen Adventuring Blog.

YOUR KNOWLEDGE HAS VALUE

- We will publish your bachelor's and
  master's thesis, essays and papers

- Your own eBook and book -
  sold worldwide in all relevant shops

- Earn money with each sale

Upload your text at www.GRIN.com
and publish for free